Wojciech Niewęgłowski • Łukas

BMW R 75

and other BMW motorcycles in the German Army 1930–1945

KAGERO

The column of *BMW R 12* motorcycles in front of the Hradčany Castle in Prague, spring 1939. There are two types of fenders installed in this popular inter-war era model visible (Bundesarchiv).

Bavarian rotating propellers

The story of one of the modern automotive giants is dated shortly before the outbreak of the Great War. In 1910, in southern Germany two manufacturing plants were set up, widely regarded as the forerunner of the official *BMW*: *Aeroplan Otto-Alberti*, and three years later: *Rapp Motoren Werke GmbH*. Both have taken mounting aircraft engines, the first of the companies a few months before the Sarajevo events transformed into *Otto Flugzeugwerke* signed a contract for the construction of airplanes for the Bavarian Army. While the imperial troops fought in the trenches, in case of both manufacturers important transformation took place: in 1916 owner of *Bayerische Flugzeugwerke GmbH* – Gustav Otto stood on the brink of bankruptcy, and in two consecutive years, Karl Rapp has changed its company to *Bayerische Motoren Werke AG*, soon selling its first shares. The new institution, the head of the board was Franz Josef Popp and chief engineer - an engineer Max Friz received from the German authorities' request for two thousand airplane engines. This impulse lay at the basis of dynamic development and the related success, and above all – the operational ceiling record set in June 1919 using the *BMW IIIa* engine (9.760 m). Around the same time, there were sort of a financial revolution – part of the shares acquired one of the richest men in contemporary Europe – Camillo Castiglioni who in a short time, passed the majority

to the Association of Banks in Vienna. As a result of the transfer was carried out in 1922, the financier merged with companies formerly managed by Gustav Otto and became the real owner of the rights and technology, *BMW*, this condition has persisted for the next seven years, until the takeover by a consortium of German banks. This circumstances allowed the Bavarian plants survive the crisis of the first postwar months especially that conducted before the activity is interrupted as a result of limiting the Armed Forces of the Weimar Republic by the preliminaries of the Versailles Treaty[1].

The first two-wheel vehicles under the symbol if tge blue and white rotating propeller[2] was presented nineteen years after the first motorcycles had appeared in the army of the Hohenzollern Empire[3]. Financially secure company, trying to take advantage of favorable conditions for the construction of this type of machines, caused by limitations imposed on Germany at the Versailles, first proposed *M2B15 BMW* engine regarded as the universal one and finally in 1923, presented its own proposal in the medium class motorcycle – *BMW R 32*[4]. The design of the equipped with two-cylinder *M2B33* engine with a capacity of 486 cm^3 and power of 8.5 HP the engineer-duo of Max Friz abd Martin Stolle created in just five weeks. The trademark of the machine, which forms the peculiar symbol of the brand in the next decades were the opposite arrangement of cylinders (the boxer type engine) and the cardan shaft. In addition, the vehicle was characterized by solid, double frame, rigid rear wheel suspension and a triangular, flattened tank with a capacity of fourteen liters. The machine weight was 122 kg was able to reach a maximum speed of 95 road km/h. Average fuel consumption by optimum driving was 3 liters per 100 km. If necessary, the sidecar could have been attached. The motorcycle was produced for three years, the number of units sold amounted to three thousand and ninety.

Less then fourteen months had passed since the September debut of the *R 32* at the Motor Show in Berlin, where another motorcycle rolled off the Munich factory again. *BMW R 37* was equipped with a drive unit with a slightly increased capacity (494 cm^3), but nearly twice the power - 16 HP. Moreover, it was decided to use the innovation created by engineer Rudolf Schleicher – the cylinder heads equipped with aluminum covers, the patent which is used by modern times. Even before the launch of production, which lasted between 1925 and 1926, the vehicle made a loud because of its sport performance: using *R 37* Franz Bieber gained a victory in the *ADAC Eifelrennen* mountain race and the Ruselberg Rally. Soon after, another *BMW* driver – Rudi Reich led his 170-kg sidecar set on the highest step of the *Ettaler Bergrennen* podium. Taking advantage of the growing interest in 1927 in the manufacturer catalogue, the 18 HP powered *BMW R 47* was also revealed.

The results mentioned above significantly contributed to strengthen the position of *Bayerische Motoren Werke AG* in the German market. In the mid-twenties, the company of Franz Josef Popp could have stand boldly compete with manufacturers such as *NSU Motorenwerke AG* from Neckarsulm in Baden-Württemberg and the acclaimed Nuremberg *Zündapp GmbH*. The demise of the first motorcycle presented above coincides with the launch of its upgraded and above all cheaper version of *R 42*[5]. Within two years the civilian users have purchased about six and a half thousand pieces of this variant, adding another nearly fourteen hundred of model *R 52* – the successor manufactured between 1928 and 1929. Despite the signs of the great economic crisis which has not spared the German State, *BMW AG*, could boldly declare success[6].

BMW in the German Army

With the increase in demand for civilian market there were the Armed Forces officials who started to look closely at the machines with the blue-and white propeller sing. The demand increased significantly after announcing of the motorization program at the end of the twenties.

The first model, which drew the attention of the Ministry of the *Reichswehr* was mentioned earlier *BMW R 52* and the heavy *R 62*. For their use primarily spoke optimum engine power – respectively, 12 HP (486 cm^3) and 18 HP (745 cm^3) and a solid construction. These models can therefore be regarded as the precursor of the military use of two-wheel vehicles from the capital of Bavaria. Number of them appeared on the ranges, however, three years later through another medium variant – *BMW R 4*. Summer of 1935 during maneuvers in Munster for the first time German Army presented the motorcycle battalion, the basis of which were the *Zündapp K800* and *BMW R 4* models.

German military police patrol with *BMW R 12*. The pouch is probably made of canvas and the police license plate started with *Pol* letters is visible too. Eastern front, summer 1941 [Bundesarchiv].

Senior corporal of the *Propagandakompanie 691* attached to the Armoured Group Kleist photographed in the Ukraine during the early stage of the Operation Barbarossa. There is a leather pouch visible on the interior brim of the sidecar – one of the standard elements of the German military motorcycles equipment (Bundesarchiv).

BMW R 12 in the service of *Luftwaffe*. Eastern front, summer 1941. This model was produced in a record amount of nearly 38.000 exemplars; it served in the Army from 1938 [Bundesarchiv].

In the Armed Forces of the Weimar Republic, and then – the Third Reich, motorcycles have been divided into three classes, to which the assignment was dependent on the capacity of the drive unit. In each of these *Bayerische Motoren Werke AG* significantly marked their participation.

The light military motorcylces (in German: *leichte Kräder* or *Meldekräder*) represented *R 2*, *R 20* and *R 23* models, which engines did not exceed the limit of 350 cm³. The first one was produced in five series between 1931 and 1936, and despite the relatively high price – 1000 German Marks enjoyed considerable popularity, especially among people who do not have permission to drive (since 1928 driving motorcycles with engine capacity below 200 cm³ did not need a license). It was an interest in the civil sphere[7] was one of the cornerstones of the incorporation of this equipped with a single cylinder 198 cm³ and 6 HP engine (from the third series – 8 HP), to the *Reichswehr*. The vehicle, having a curved front to the front fork with a clearly visible leaf-spring suspension, holding the weight of one hundred and thirty kilograms is accelerated to a speed of 95 km / h (in the case of the 6 HP power unit – 80 km / h).

With the end of its production the company from Munich offered variant *R 20*. According to the predecessor this model was above all distinguished by a telescopic front wheel suspension, typical for this version, placed just above the axle, rubber shields as well as reduced to 192 cm³ engine capacity, having no effect on the power. Simultaneously the fuel tank was extended by two liters.

The last vehicle of the series in the inter-war era was about five kilograms heavier *BMW R 23*. The production of this variant lasted from 1938 to 1940. The vehicle was equipped with the engine with capacity of 247 cm^3 and a power of 10 HP, which allowed to reach a maximum road speed of 95 km / h.

The first *BMW* military motorcycle representing the medium class (*mittlere Kräder*) and the first under the sign of a spinning propeller in general who became interested in military officials in Germany was the *R 52* model. However, in this category company from Munich has established itself in 1932 with the produced through the next five years in five series and more than fifteen thousand copies, *R 4* variant. The vehicle, originally considered as the primary purpose two-wheel vehicle of the army, was equipped with a single cylinder, OHV engine, having a capacity of 398 cm^3 and the power of 12 HP initially, and eventually - 14 HP. As in the earlier models was the cardan shaft drive. Also uses dry clutch and manual transmission with four gears[8]. Maximum road speed was ranked in the range of 110 km / h. The front wheel received the telescopic suspension, while the rear was rigidly mounted in a frame of pressed steel. External element that changed during the course of completing another sampling was front fender, initially (until the third series) with a shape similar to a lighter *BMW R 2*. The curb weight of the machine was one hundred and twenty kilograms. Because of its almost legendary reliability, the *BMW R 4* found widespread use in both the civilian and military space, being used until the collapse of the Third Reich in training centers or by paramilitary organizations.

BMW R 12 Gespann **leading the military column during the French Campaign of 1940. There is a characteristic for a part of the producion one-piece front fender clearly visible [Bundesarchiv].**

German motorcycle troops during the rest somewhere in the USSR in summer 1941. The construction details suggest that the photographed vehicles represent *BMW R 12* type. The unit visible from the rear is equipped with both leather pouches, which were added to nearly every German military motorcycles [Bundesarchiv].

SS *Totenkopf* **Division soldiers during the Operation** *Barbarossa.* **There is** *Bayerische Motoren Werke AG* **emblem visible on the** *BMW R 12* **frame – the blue-and-white spinning propeller [Bundesarchiv].**

Despite the good performance both on road and in the field, the *R 4* model with the time he began to give way line in units of his successor. Produced between 1937 and 1940 *BMW R 35*, sometimes called *The Donkey* (*Esel*), proved to be the most popular solo motorcycle, ever used by the German army. This one hundred fifty-five kilograms vehicle was propelled by 14 HP, 340 cm³ engine, obtaining on the highest, fourth postpone-

ment the speed of 100 km/h. With a balanced traveling at 60 km/h the fuel combustion finished in the borders of three liters per one hundred kilometers, which meant that the vehicle can be regarded as a very economical (the maximum road range was approximately four hundred kilometers). It is worth noting that as the only of all bikes offered by *Bayerische Motoren Werke AG* during the Weimar Republic and the Adolf Hitler,

BMW R 75 Gespann **manufactured probably in 1942, attached to the Hermann Göring Brigade. There are metal front bumper covers and coloured emblem fastened on the fuel tank side clearly visible. The sidecar is armed with** *7,92 mm Maschinengewehr 34* **[Bundesarchiv].**

was also complete by the end of the World War II, the production in the factory of Eisenach, was continued until 1955.

Although the there was the *Zündapp* of Nuremberg which clearly dominated in the category of heavy motorcycles (*schwere Kräder*), the *BMW* proposed most of its two-wheel vehicles strictly in that group. Can not be ruled out that this fact was associated with the demand not only reported by the military but also civilian users, for which the heavy sidecar unit, was not too expensive, and relatively capacious and durable mean of transportation especially in the eve of great economic crisis and subsequent years. At the turn of the third and fourth decade of the twentieth century, the armed forces have stated so clearly the requirements for a set of machines equipped with engines with a capacity exceeding 500 cm³. Due to the anticipated use in a transportation role for the infantry in direct combat area, as well as poorly developed road network in Eastern Europe – the main direction of the planned German expansion, indicated that the ground clearance of the bike should be no less than 100 mm below the vehicle and 250 mm in a case of a sidecar.

In 1928, the Ministry of the *Reichswehr* and the Board of *BMW AG* entered into an agreement under which over the next several months, the Army planned to provide reliable, heavy motorcycle. There was the R 62 model which were supposed to fulfill the army demand, with a two-cylinder 745 cm³ and 18 HP engine, being allowed to reach a maximum road speed of up 115 km/h.

Especially for use with a sidecar its successor was designed – *BMW R 11*, produced in five series between 1929 and 1934. Although already a successful debut at the Olympic Games in London in 1928, the military use of that variant was rather a temporary solution. The machine was equipped with 745 cm³ and 18 HP engine (the last production series showed a 20 HP). Three-speed transmission with dry single-plate clutch (since 1931 – double-plate) allowed to disperse the more than 160 kg vehicle to the top speed of 112 km/h Noteworthy is also the fact that the variety is described as the first in the history of motorcycles under the sign of a spinning propeller had a frame made of pressed steel sections and was equipped in standard with traffic lights and speedometer[9].

Approximated performance was characteristic for another model proposed to the Third Reich Army – *BMW R 12*, which was developed in the mid-thirties as an answer to the needs of a motor tourism increasingly popular in Western Europe. In the four years of production (1938-1941), the number of completed and sold exemplars could have reached even 38.000. The engine of this variant was a twin-cylinder 745 cm³ and 18 HP power unit with a single carburetor in an early version j (more popular in the army) or 20 HP in the case of the model with two. There were dry, double-plate clutch, and four hand-selected gear ratios installed too. The

BMW R 75 Gespann **in the German paratroopers service, Italy 1943 or 1944. The unit is armed with** *7,92 mm Maschinengewehr 42* **[Bundesarchiv].**

motorcycle reached a maximum speed of 110 km/h (single-carburetor model) and 120 km/h (double carburetor model). After attaching the sidecar it decreased to 85 km/h. The vehicle also had a number of innovative solutions, including in particular the depreciation of the telescopic front wheel with damping oil (the first of its kind in the world) and, as was the exception to the rule, rear drum brakes (so far was the most common variant of the jaws affecting the shaft). The confirmation of its popularity may be the fact that after 1942 it was one of the two (along with the *BMW R 75*) motorcycles bearing the symbol of a spinning propeller, manufactured for the Armed Forces of the Third Reich by the company from Munich.

Also another heavy motorcycle designed in Munich was bearing the name of the *Wehrmachtsmodell* or *Wehrmachtsk-rad*. Over 180 kg *BMW R 61* has been designed nearly almost exclusively for a sidecar. The two-cylinder engine with a capacity of 587 cm³ and the power of 18 HP dispersed that variant to a speed of 100 km/h, and in the solo version – 115 km/h.

BMW R 75 Gespann **during the spring thaw of 1943. There is a right pouch mounting and the canvas sidecar cover clearly visible [Bundesarchiv].**

Deutsches Afrikakorps **despatch riders resting with their** *BMW R-12s*. **Libya, April 1941 [Bundesarchiv].**

A year before the outbreak of the World War II, assembly halls in Munich began to leave the next version – *BMW R 71* produced in parallel to the intended mainly for the civilian market variants: *R 61*, and sport *R 51*. This, finalized in the amount of three and a half thousand pieces vehicle was equipped with a two-cylinder 746 cm³ and 22 HP engine. Curb weight in a solo variant was 190 kg, while the sidecar – about 300 kg. The vehicle accelerated, respectively to 125 km/h and 95 km/h.

The Sahara

Development of the *BMW R 71* has opened the way to build a new, dedicated exclusively to the army two-wheel vehicle. Start of work came in 1937 when the competition for a heavy motorcycle with a military purpose was announced.

According to the guidelines the unit ought to be able to be connected with a machine gun equipped sidecar, the minimum ground clearance was no less than 150 mm and 16" wheel with off-road tires with the ability to install anti-skid chains in standard. According to the expectations the vehicle should transport three soldiers equipped with a complete, individual combat equipment or load of about five hundred pounds. So loaded had to disperse to 80 km/h, while providing the possibility of sustainable speeds no greater than 3 km/h. Apart from the Munich company the competition joined two other producers, founded nine years earlier *DKW* factory and respected in the automotive market in Germany – *Zündapp*. It is the latter project, designated as *KS 750* proved to be most appropriate. In 1940 it was decided to implement the concept, while putting emphasis on the management of *BMW AG*, in the shortest possible time began to produce

Table 1. *BMW* light motorcycles used by the German Army 1930-1945 – basic technical data			
	BMW R 2	**BMW R 20**	**BMW R 23**
Engine type	four-tact OHV boxer	four-tact OHV boxer	four-tact OHV boxer
Drive	cardan shaft	cardan shaft	cardan shaft
Cylinders	1	1	1
Bore	64 mm	64 mm	68 mm
Engine capacity	198 cm³	192 cm³	247 cm³
Engine power	6 or 8 HP	8 HP	10 HP
Maximum speed	95 km/h	95 km/h	95 km/h
Fuel tank capacity	10 l	12 l	10 l
Wheelbase	1303 mm	1330 mm	1330 mm
Weight	130 kg	130 kg	135 kg

BMW R 12 Gespann captured by the British soldiers in the Tobruk area, December 1941. There is a big *Deutsches Afrikakorps* insignia painted on the sidecar side and the white stripes allow to indicate the vehicle in the dark (Imperial War Museum Archive).

Table 2. *BMW* medium motorcycles used by the German Army 1930-1945 – basic technical data

	BMW R 52	BMW R 4	BMW R 35
Engine type	four-tact OHV boxer	four-tact OHV boxer	four-tact OHV boxer
Drive	cardan shaft	cardan shaft	cardan shaft
Cylinders	2	1	1
Bore	78 mm	84 mm	84 mm
Engine capacity	486 cm^3	398 cm^3	340 cm^3
Engine power	12 HP	14 HP	14 HP
Maximum speed	100 km/h	110 km/h	100 km/h
Fuel tank capacity	12,5 l	12 l	12 l
Wheelbase	2100 mm	1320 mm	1300 mm
Weight	152 kg	120 kg	150 kg

that variant under the license. Franz Josef Popp and other members of management have given a negative reply, but managed to work out a compromise involving the unification of seventy percent of the finally chosen machine and the components of a competitive *BMW R 75*. In 1942 it was decided, moreover, that the production of the last mentioned model will be ended after assembling the twenty thousand and two hundred exemplar[10].

Manufacturing the motorcycle, which is known as the *BMW Sahara* started in July 1941 (first copy had a frame number 750.000)[11]. During the first months the vehicles were produced in the halls of the factory in Munich, to finally move to Eisenach in Thuringia. By the autumn of 1944, when, in connection with the damage constituting the effect of Allied bombing raids the production was ceased, there were sixteen thousand five hundred forty-five pieces completed. Another

Table 3. *BMW* heavy motorcycles used by the German Army 1930-1945 – basic technical data

	R 62	R 11	R 12	R 61	R 71
Engine type	four-tact SV boxer	four-tact SV boxer	four-tact SV boxer	four-tact SV boxer	four-tact SV boxer
Drive type	cardan shaft	cardan shaft	cardan shaft	cardan shaft	cardan shaft
Cylinders	2	2	2	2	2
Bore	78 mm	78 mm	78 mm	78 mm	78 mm
Engine capacity	745 cm^3	745 cm^3	745 cm^3	597 cm^3	746 cm^3
Engine power	18 HP	18 HP	18 or 20 HP	18 HP	22 HP
Maximum speed	115 km/h	100 km/h	110 lub 120 km/h	115 km/h	125 km/h
Fuel tank capacity	12,5 l	14 l	14 l	14 l	14 l
Wheelbase	14000 mm	1380 mm	1380 mm	1400 mm	1400 mm
Weight	155 kg	162 kg	185 kg	184 kg	190 kg

Tabela 4. *BMW R 75* –maximum speed of the individual gears				
Road gear	1	2	3	4
Speed	22 km/h	44 km/h	66 km/h	92 km/h
Off-road gear	1	2	3	4
Speed	14 km/h	24 km/h	42 km/h	-

Reverse — 4 km/h

German 5th Armoured Regiment dispatch riders during the railway transport in summer 1940. Both of the vehicles represent the *BMW R 12 Gespann* type [Bundesarchiv].

ninety-six left the factory between 1945 and 1946, already under Soviet auspices.

In the course of the production the *R 75* variant was extensively renovated. The most important ones were: strengthening the foot brake lever, differential lock and transmission springs. In addition, bold tube frame, increased ground clearance and front fender were narrowed (later the width of the rear equivalent was also reduced). At the end of the second year of the war against the Soviet Union there was also air filter modified by increasing its size, adding a felt pad and moving the whole right side of the top surface of the fuel tank in place of a shallow rectangular tray. The whole was covered with a round bonnet, opened in the direction of the gear lever. The driver also received a front leg shield plate. Exemplars with frame of 762.260 and over were deprived of metal casings shock front – in their place a black rubber covers were installed. The oil deflector was simplified and the mounting of the passenger footrests changed its location too.

The engine was an air-cooled, two-cylinder, overhead valve, four-stroke *BMW M5 BM 275* with a capacity of 746 cm^3 and power of 25HP. The transfer of power to the rear wheel as standard ensures cardan shaft. The wet air filter and two carburetors type *Sa 24Gräzin* were installed. Not wanting to take the risk of battery failure, especially in common front conditions, the magneto *Norris ZG2a* or *Bosch FJ 2R 134* were chosen. In addition the dry, single-plate clutch with power comparable to the assembled counterparts in medium-size civilian cars, and also allows the manual or foot selection of four road gears, one reverse or three off-road. In theory, there was a choice of the fourth off-road, but this operation have had to disconnect the drive; from the exemplar with frame number of 758.015 this problem was solved. At the gearbox exit, rubber bumpers and the cross-joint were installed to prevent an hardness of selection. Knob located on the right wall of the fuel tank at the same time served as an gear indicator. The maximum speed of the unit was 92 km/h.

Full tank, with a capacity of twenty-four liters (three liters reserve) provide the maximum range about three hundred and fifty kiolmeters[12]. Average fuel consumption during the road driving was therefore about 7 l/100 km, while in the field – 9 l/100 km. The oil tank once housed two liters of fluid. As the motorcycle devoted primarily to be connected with a sidecar, as in the case of *Zündapp KS 750*, this element has been also driven. By using locked differential the power distribution was 70% for the rear motorcycle wheel and 30% for the wheel of the side trailer. It also received a separate hydraulic brake which could have been disconnected automatically at the moment of taking away the sidecar. In an innovative way the construction of the welded steel tube frame had been prepared – using some of its segments, in the event of a damage disassemble the whole was not needed[13]. The front fork is equipped with a massive, double filled telescopic oil

shock absorber. 16" wheel with off-road tires have received a life span of more than four times the theoretical standard versions mounted in typical motorcycles. The air pressure was 1.75 atm. for the front and the trolley and 2.75 atm. for the rear. *BMW R 75 Gespann* weight reached the four hundred and twenty pounds. Almost the same heavy load he was able to take on board. The spare wheel was installed on the trailer behind the passenger place; the tool kit, headlight cover and large, leather pouch or metal tray mounted on the left side below the rear seat were also added.

There was essentially one type of a sidecar provided for the famous *Sahara*, consisting of *Steib W Krad B2* nacelle mounted on thick-walled tubes of the *Beiwagen 40* or *Beiwagen 43* chassis. The relative comfort of the passenger provide rocker wheels and a pair of rear half-springs and rubber sleeves at the front. In standard version the sidecar was equipped with *7,92 mm Maschinengewehr 34* or *42* mount, and two symmetrically arranged near the front edge of the rectangular leather pouches or metal stowage bins. Most exemplars had the rear tow bar. First models were fully equipped with road lights placed on the fender, but they were soon abandoned.

The combat debut of the *BMW R 75* took place during the war in North Africa and the Eastern Front, in the summer of 1941. Because of its off-road capability in a short time machine earned the favor of the members of all branches of the Third Reich Army.

Fording of the *BMW R 12 Gespann* unit somewhere in the ekstern front, summer 1941. There is *Bayerische Motoren Werke AG* emblem visible on the sidecar side [Bundesarchiv].

Epilogue

The *BMW* motorcycles, in addition to competing *Zündapp* ones are one of the symbols of the German automotive industry of World War II and the inter-war era. The most recognizable two-wheel vehicles under the sign of a spinning propeller is without doubt the *R 75* model, readily used not only by German soldiers but also their opponents, even after the capitulation of the Third Reich. His versatile and durable design had a partial effect on the variants built by *Bayerische Motoren Werke* plant in the next decades, such as *R 51/3* (production in the years 1951-1954), *R 67* (1951), or *R 68* (years 1952-1954). In its original form the Eisenach factory attempted in the mid-fifties to reactivate production under the designation *AWO 700*, but the idea has not been finally realized. A copy of the *Sahara* was also made by the *US Army*

German dispatch rider poses with *BMW R 35* popularly called *The Donkey*. This type of motorcycle, which first exemplars were sent to *Wehrmacht* in 1937, was produced by the Eisenach factory even ten years after the end of the World War II [Bundesarchiv].

Table 5. *BMW R 75* – basic technical data

	BMW R 75
Engine type	four-tact OHV boxer
Rodzaj napędu	cardan shaft
Cylinders	2
Bore	78 mm
Engine capacity	746 cm^3
Engine power	26 HP
Maximum speed	92 km/h
Fuel tank capacity	24 l
Maximum road range	350 km
Wheelbase	1444 mm
Dimensions with sidecar	
Length	2430 mm
Width	1730 mm
Height	1000 mm
Ground clearance	150 mm – 270 mm
Weight with sidecar	420 kg

(experimental *Harley Davidson XA*) and the Soviets (*TMZ-53*). Whatever the model, the *BMW* brand is a mark of quality even in the modern times, the quality which evolved during the the last global conflict.

Bibliography

Books and articles:

"Osiołek" czyli małe BMW, http://pulk37.org/2012/06/05/osiolek-czyli-male-bmw/ (data dostępu: 5.06.2012 r.).

BMW – Bavaria's Driving Machines, Nowy Jork 1984.

Handbuch zur Anleitung, Wartung und Bedienung der BMW-Zweizylinder-Modelle, Monachium 1925.

Birch Gavin, *Motorcycles at War*, Barnsley 2006.

Dugan Ken, *Kradschützen and Kradmelder at the Front*, [w:] "Allied-Axis. The Photo Journal of the Second World War", No 11, Delrey Beach 2003.

Dugan Ken, *Kradschützen and Kradmelder at the Front*, [w:] "Allied-Axis. The Photo Journal of the Second World War", No 18, Delrey Beach 2006.

Hinrichsen Horst, *Kräder der Wehrmacht. Ausbildung und Einsatz 1935 bis 1945*, Wölfersheim-Berstadt 1997.

Hinrichsen Horst, *Schwere Beiwagenkräder der Wehrmacht 1935-45*, Wölfersheim-Berstadt 1999.

Hoppe Henry, *BMW Motorcycles of the Reichswehr and Wehrmacht 1928-1945*, [w:] "German Military Vehicle Rarities (1). Imperial Army, Reichswehr and Whermacht 1914-1945", Erlangen 2004.

Knittel Stefan, *Deutsche Kräder im Kriege. BMW, DKW, NSU, Triumph, Vitktoria, Zündapp*, Dorheim 1982.

Krzal Franz, *Ein Motorrad der Vergangenheit*, http://www.lookover.at/Berichte/Natur_Technik/2007/EIN_MOTORRAD_DER_VERGANGEN.php (data dostępu: 5.06.2012 r.).

Lepage Jean-Denis, *German Military Vehicles of World War II*, Jefferson 2007.

Linz Harald, *Wielka encyklopedia samochodów*, Łódź 1992.

Stünkel Udo, *BMW-Motorräder Typenkunde : Alle Serienmodelle ab 1923*, Bielefeld 2008.

Wilson Hugo, *The Encyclopedia of Motorcycle*, Londyn 1995.

Websites:

BMW Geschichte, BMW Group Archiv, www.historischesarchiv.bmw.de.

BMW Klub Warszawa, www.bmwklubwarszawa.pl

MBike, www.mbike.com

Endnotes

[1] There had even a project occured to transform the BMW facilities so that they could have undertaken the manufacturing of furniture and kitchenware.

[2] As a symbol of *Bayerische Motoren Werke AG* a round shield divided into four, arranged alternately white and blue fields - the national colors of the Kingdom of Bavaria was adopted, symbolizing the spinning propeller. This manufacturer's insignia is still in use.

[3] The first machine of this type revealed during maneuvers of the imperial army, organized in 1904. Recognized the advantages of a motorcycle as a cheap and easy to use means of transport as a result, during the First World War the German army had been equipped with five thousand four hundred two-wheel vehicles.

[4] In the early twenties *Bayerische Flugzeugwerke AG* effort was a series of motorcycles bearing the names: *Helios* and *Flink*, with *M2B15* engines installed.

[5] According to prices in the mid-twenties, for the *BMW R 32* motorcycle it had to be paid 2200 German Marks, while for the *R 42* version – 1500.

[6] The success was grounded because of the highly popular, produced from 1927 small car *BMW Dixi* - first car offered by the Bavarian company.

[7] There were 15.000 exemplars sold in the first half of the thirties.

[8] The first production series motorcycles were equipped with three gears.

[9] The development version of the *R 11* model was a *R 16* sport variant equipped with a OHV engine with a capacity of 735 cm^3, which was able to achieve maximum speed up to 145 km/h. With that kind of vehicle the team composed of Ernst Henne, Ludwig Kraus, Josef Sepp Mauermayer and Sepp Stelzer won the XV Six-Day Rally in Wales.

[10] At the same time the idea of creating a hybrid of *BMW* and *Zündapp* motorcycle called *BW 43* arose, which however did not go beyond the conceptual stage.

[11] It was planned to complete twenty five exemplars daily.

[12] There was an manufacturer emblem installed on the fuel tank sides either in a separately made, coloured version or the pressed-in one.

[13] As Tadeusz Pawlak stated in the subunit of ten to twelve motorcycles tere were three spare BMW frames present, what allowed to make the field repairs faster.

BMW R 75 and other BMW motorcycles in the German Army 1930–1945
Wojciech Niewęgłowski • Łukasz Gładysiak
First edition • LUBLIN 2012 • ISBN 978-83-62878-39-0

© All rights reserved.
• Translation: **Łukasz Gładysiak** • 3D profiles: **Wojciech Niewęgłowski** • Archive photos: **Bundesarchiv, S. Knittel, H. Hope, Imperial War Museum Archive,**
Today's photos of *BMW R 12 Gespann*: **Grzegorz and Michał Buczek** • Design: **KAGERO STUDIO**

Oficyna Wydawnicza KAGERO • www.kagero.pl • e-mail: kagero@kagero.pl, marketing@kagero.pl

Editorial office, Marketing, Distribution: OW KAGERO, Akacjowa 100, Turka, 20-258 Lublin 62, Poland, phone/fax: (+48) 81 501 21 05

BMW R 12 Gespann photographed probably during the first winter on the eastern front. Besides the *R 75* model it was the only heavy variant manufactured by the Bavarian factory for the German Army after 1942 [Bundesarchiv].

Soldiers of the reconnaissance battalion of the German 1st Armoured Division phtographed in the Leningrad area in summer 1941. *BMW R 12 Gespann* is covered with the complete tactical insignia set painted on the front fender [Bundesarchiv].

German dispatch rider with *BMW R 4* photographed in the Florence area in 1944. There are characteristic suspension elements including a single shock absorber clearly visible [Bundesarchiv].

BMW R 12 technical overview. The civilian, Berlin license plate suggests that the machine was incorporated by the Army during the pre-war requisitions. The demand for the two-wheel vehicles under the sign of the spinning propeller which was rising from the beginning of the thirties allow the armed forces to secure the motorcycles supply also from the secondary sources [Bundesarchiv].

Brand new *BMW R 12 Gespann* units during the railway transport to the eastern front. Autumn of 1941 [Bundesarchiv].

BMW R 75 equipped with metal front bumper covers and the air filter mounted on the fuel tank. There is an exhaust muffler characteristic for this model visible below the passengers seat [Foto via S. Knittel].

BMW R 75 equipped with rubber front bumper covers, characteristic for the exemplar with the 762.260 and the following frame numbers [Foto via S. Knittel].

The final model of the *BMW R 75* air filter, installed on the upper side of the fuel tank [Foto via S. Knittel].

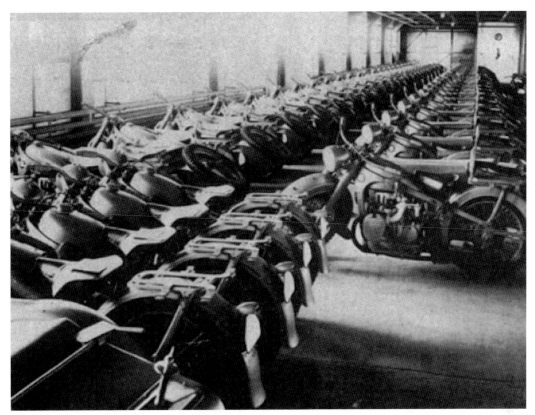

BMW R 12 solo and sidecar motorcycles waiting for their military recipients. This model was one of the most common manufactured by *Bayerische Motoren Werke* AG ntil the end of the World War II [Foto via S. Knittel].

Luftwaffe dispatch riders posing with *BMW R 35*. Germany, the turn of the thirties and fourties [Foto via H. Hope].

BMW R 35 – the medium motorcycle with the spinning propeller emblem used by the German Army since 1937. There is a sump cover visible on the photograph, an element installed in the military-purpose models [Foto via H. Hope].

BMW R 11, the model manufactured inter alia with a vie to the *Reichswehr* units, in the service of the Third Reich Army [Foto via H. Hope].

BMW R 35 during the eastern front campaign of 1943 [Foto via H. Hope].

BMW R 12 Gespann of Mr. Grzegorz and Mr. Michał Buczek, members of the AA7 Reenactment Association (Poland). The motorcycle was produced in 1938.

The author wishes to thank the vehicle Owners for the opportunity to use the pictures for the purposes of this publication.

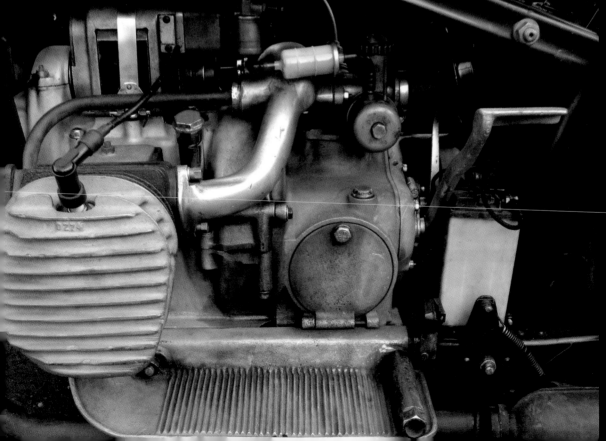

Krad B

Gesamt-Höchstbelastung 220 kg
Belastung vorn ein-
schließlich Insasse 170 kg
Gepäckraumbelastung ein-
schließlich Ersatzrad 50 kg

BMW R 75 Gespann in six views. The motorcycle is equipped with sidecar and metal transport cases. The armament consists of 7,92 mm Maschinengewehr 34 light machine gun.

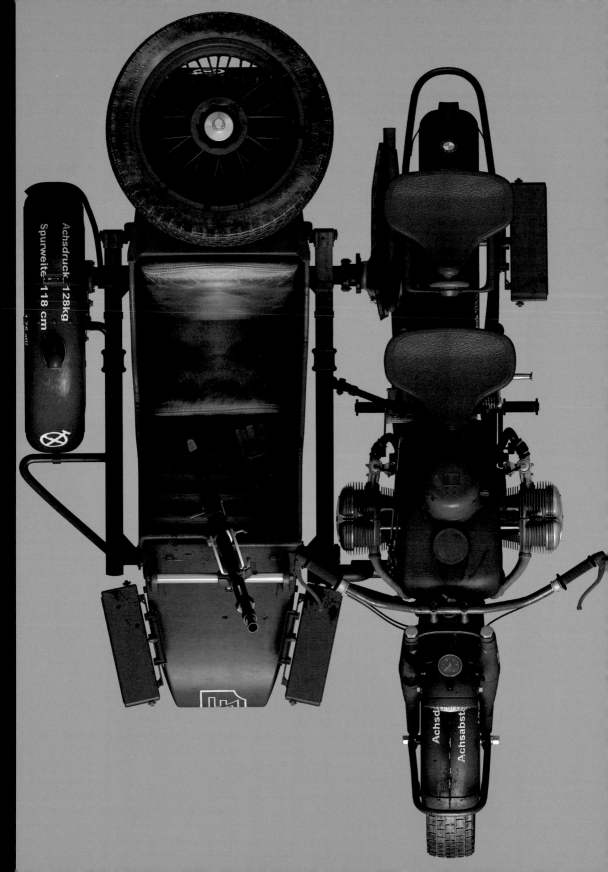

BMW R 75 Gespann of the SS Panzergrenadier-Division Das Reich motorcycle battalion, eastern front, autumn 1942. The monotone dark grey camouflage had been applied as well as the

ont view of *BMW R 75 Gespann*. There are metal front bumper covers and transport cases on the brims of the sidecar visible. The
ont light blackout cover had been removed.

ght front view of the *BMW R 75 Gespann*. There are white exploitation markings and the *Wolfsangel* runic *SS Panzergrenadier Divi*

Left view of BMW R 75 Germany. The pressed Bayerische Motoren Werke AG emblem and the air filter with its large cover installed on the upper part of the fuel tank, drive attention

ar view of *BMW R 75 Gespann*. There is the tactical marking of the motorcycle battalion painted in white on the sidecar deck.

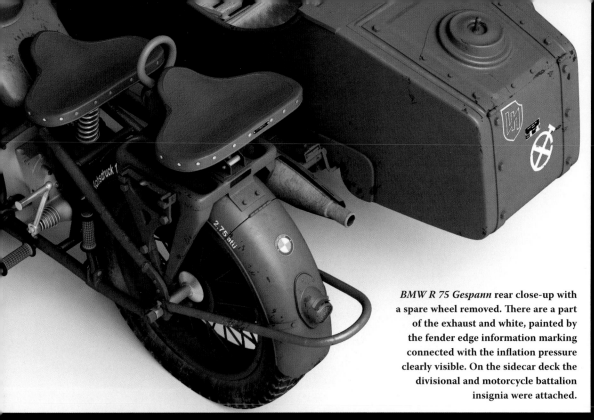

BMW R 75 Gespann rear close-up with a spare wheel removed. There are a part of the exhaust and white, painted by the fender edge information marking connected with the inflation pressure clearly visible. On the sidecar deck the divisional and motorcycle battalion insignia were attached.

Steib W Krad B2 sidecar dedicated to *BMW R 75* motorcycles. The passengers tubular bracket, transport case and machine gun mounting drive attention. There is also a road light on the fender, an element which was characteristic especially for the early production series.

Left view of *BMW R 75*. There was an exhaust with its characteristic cover installed high, just below the passenger seat. This solution prevented the attachment of the second pouch or transport case.

BMW M5 BM 275 engine close up. This unit, with a capacity of 746 cm3 and 25 HP was installed in the *BMW R 75* motorcycles.

The center section of *BMW R 75* close-up. There is an air filter with its characteristic cover installed on the upper part of the fuel tank with the rubber pads increasing driver comfort on the sides. The speedometer is an element of the front road light.

BMW R 75 Gespann after removing the sidecar gondola. There are white exploitation markings visible on the fenders.

BMW R 75 Gespann sidecar chassis close-up. There are half-springs of the gondola clearly visible as well as the fender mount.

Upper view of the *BMW R 75 Gespann* sidecar fender. The mounting system, gondola half-spring and the tubular leg support for the passenger are clearly visible.

BMW R 75 front disc brake and fork mounting detail.

BMW R 75 front light with integrated speedometer. The handlebar mountings drive attention too.

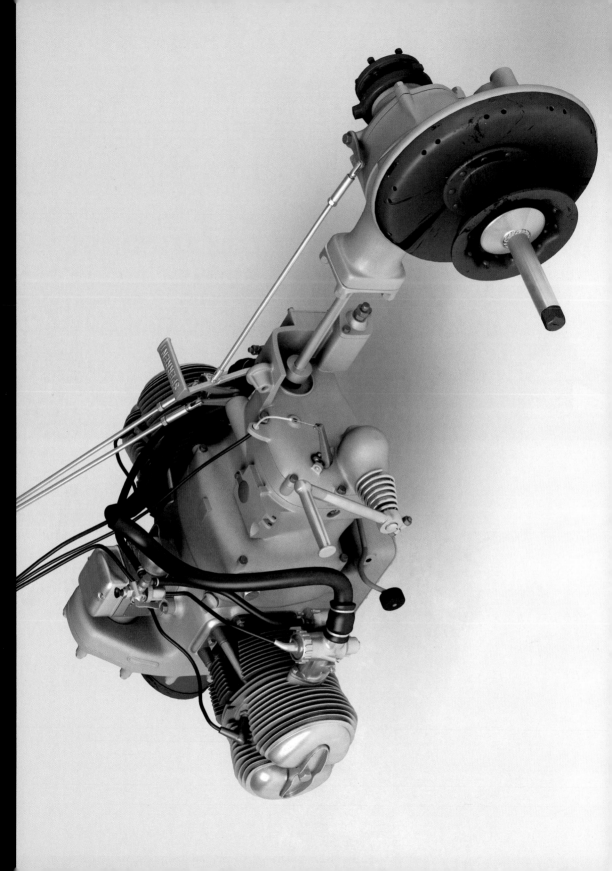

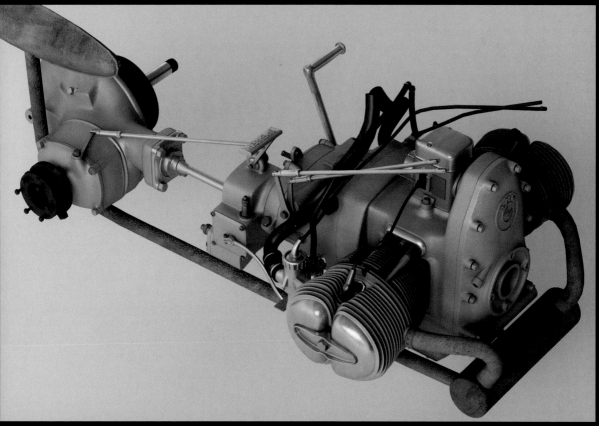

BMW M5 BM 275 engine of the *BMW R 75* motorcycle with the exhaust system installed.

BMW R 75 Gespann of the German 15th Armoured Division motorcycle battalion, North Africa, 1942/1943. The vehicle was repainted with sand-yellow monotone camouflage applied on top of the original dark grey one. There is a twenty liters water jerry can attached on the spare wheel and the 7.92 mm *Maschinengewehr 42* without a mount inside the sidecar.

Rear view of the *BMW R 75 Gespann* of the German 15th Armoured Division. The motorcycle seats and round passenger handle are clearly visible as well as the rear fender details with manufacturer emblem in full-color version. There is a 20-liter water jerry can attached on the spare wheel.

BMW R 75 Gespann sidecar details with wooden floor planks, 7.92 *Maschinengewehr 42* and the ammunition box visible.

Upper view of the *BMW R 75 Gespann* sidecar. The metal transport cases mountings and front passenger handle drive attention.

BMW R 75 handlebar with front light without blackout cover and the front part of the fuel tank. There is the *Deutsches Afrikakrops* symbol painted in white on its side. The air-filter metal cover is also present.

BMW R 75 Gespann of the unknown *Wehrmacht* unit. Normandy, August 1944.

BMW R 75 Gespann of the unknown *Luftwaffe* unit. Normandy, July 1944.

BMW R 75 Gespann from the unknown *Wehrmacht* unit. Eastern front, December 1943.

New Releases October 2012

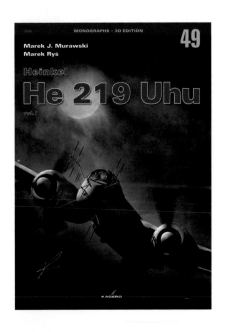

MONOGRAPHS · 3D EDITION

49

Marek J. Murawski
Marek Rys

Heinkel
He 219 Uhu

vol. I

KAGERO

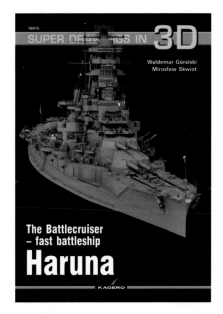

SUPER DRAWINGS IN **3D**

Waldemar Góralski
Mirosław Skwiot

The Battlecruiser
– fast battleship
Haruna

KAGERO

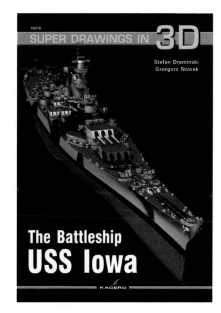

SUPER DRAWINGS IN **3D**

Stefan Draminski
Grzegorz Nowak

The Battleship
USS Iowa

KAGERO

No.2

SUPERmodel

INTERNATIONAL

KAGERO PUBLISHING

JAGDPANTHER — DML 1:35

SU-122-54 — Scratchbuild 1:35

Master Box 1:35
BMW R75

T-55 — Tamiya 1:35

POLISH STEEL

MASTER WORKSHOPS · ADVANCED PAINTING & WEATHERING TECHNIQUES

BMW R75